Lancaster's Road
The Historic
Columbia River Scenic Highway

by
Oral Bullard

TMS Book Service
P.O. Box 1504
Beaverton, Oregon 97075

I.S.B.N. No. 0-911518-64-9

This book is dedicated to all who love the Columbia Gorge, and more specifically, to those pioneer road builders who cared enough to spare it from devastation.

For the traveller who wishes to journey along the 2 major sections of the Scenic Highway that remain open to automobile traffic, those eastbound from Portland on Interstate 84 will find Columbia River Scenic Highway signs beginning as they near Troutdale and the Sandy River. There are several opportunities to turn off the freeway and on to the Scenic Highway. For the Rowena Crest section, however, the turn off is at Mosier, and it is clearly marked by a Scenic Highway sign.

Westbound travellers on Interstate 84 will find these same turn off designations, one near Rowena, just west of The Dalles and the Ainsworth Park turn off between Bonneville Dam and Multnomah Falls.

Dedication ceremonies for the Scenic Highway on Crown Point, June 7, 1916.

contents

prologue

It was known as the king of roads. It was called a poem in stone.

The Columbia River Scenic Highway, built over a two-year period, 1913-1915, was an engineering triumph that blended a work of art into one of Nature's major masterpieces, the Columbia River Gorge.

Such happenings occur but rarely, and surely never by chance. A certain magic is required and that magic is attained by a meeting and meshing of minds, talents and dreams in order to create a design worthy of the subject, and once the project is started it must be nursed through to completion, sometimes overcoming immense difficulties in the process.

In this atmosphere there can be no room for scoffers. When the course is set, positive is the watchword. Visions can become reality only through remarkable efforts and dedication.

It seems, now, that the highway was built in gentler, less fretful times. Yet that may be only present day imagination or yearning. The facts are that while it was under construction a major war broke out in Europe and shortly after it was completed the United States was involved in that war. It is known that the military establishment had long pressured for a road to the coast from eastern Oregon, and while the Scenic Highway was being built another section of the same highway was pushed through to the Pacific.

Let us put such practical considerations for roadbuilding aside for a moment and think instead of Camelot, of castles on the Rhine and things more romantic than wars. Practicality may have spurred the road's creation, but its eventual shape was the result of a vision, and this is the story of a dream realized. It is the story of a group of men who decided to do, if not the impossible, at least the difficult and improbable. In the age of realism they built a phantasy and they did it in an incredibly short time and, by today's standards, at an equally incredible low cost.

No one man can be given full credit for building the highway. In the many minds that meshed to form this particular ideal and insure its reality there was a rare mixture of talents peculiarly suited for each task.

Samuel Hill, wealthy lover of natural beauty, student of Continental roadbuilders, was past-president of the American Roadbuilders Association and given to dreams that were not just idle fancies. A frequent traveller to Europe, he had studied and admired their finer highways and, feeling his own fame rested on promoting beautiful and excellent quality roads, was determined to push a unique and singular highway through the Columbia River Gorge.

Simon Benson, another man of wealth, understood the need for the road and stirred interest in it and support for it through his many efforts and generous donations. Among the latter was the purchase of and gift to Multnomah County of three hundred acres, including Multnomah Falls and Wahkeena Falls.

John B. Yeon, the "millionaire roadmaster," was one of Portland's leading citizens. He gave more than two years of his time, without remuneration, and was in active charge of all the road construction except the bridges and hard surfacing. Yeon's particular talent for organization aided greatly in the many difficulties encountered.

Samuel Christopher Lancaster, a railroad and highway engineer of great sensitivity, had accompanied Hill to Europe to study the major highways of the continent. Lancaster, bewitched by the possibility of creating a beautiful road through a relative wilderness, was Consulting Engineer in charge of the Scenic Highway.

The roll call of names could go on for several pages. Some

others will be mentioned in the narrative that follows, some may be omitted, not by intention but through oversight.

Lancaster shared with the others a great devotion for the Gorge of the Columbia. Much of what he felt is preserved in a book he wrote, **Romance of the Gateway Through the Cascade Range**, published by the J.K. Gill Company in 1915, 1916 and 1926. Although the passage of time adds a sheen of romance to earlier efforts these writings prove that the sense of adventure and romance was prevalent at the time the highway was being thought about, planned and constructed. Certainly there were also the more prosaic considerations, but the overwhelming challenge was to satisfy the urge to build not only a road but a **beautiful** highway. That expression is used repeatedly, and in the entire project there is a simple, direct faith and belief in its ultimate success that is best expressed in Lancaster's own words.

"Our fondest dreams come true," he wrote, "if we work intelligently, earnestly, and with motives that are pure."

And so, in the end it became, uniquely, Lancaster's Road. More than any other person, his handprint is upon it. Inspired by Hill, aided by Yeon, Benson and others, this brilliant engineer with the soul of a poet crafted a road of spectacular beauty through the Columbia River Gorge, a land which he termed "this mighty work of God, done in His own way, on a scale so great than man's best efforts appear but as the work of pigmies....."

In this case, man's effort was to win national and international attention and acclaim. The dedication of the highway, in 1916, was attended by dignitaries from around the world. In 1937, more than two decades after it was completed, it was listed in a book, **Mighty Engineering Feats**, by Harriet Salt and produced by Penn Publishing Company of Philadelphia. In the book twenty-six pages are devoted to the Columbia River Scenic Highway. Other "mighty feats" with which it shares the pages are the first Continental Railroad, the Panama Canal, the New York City Water Supply, Wilson Dam, the Alaska Railroad, Boulder Dam and the San Francisco-Oakland and Golden Gate Bridges.

The advent of modern transportation — giant rigs hauling

9

huge loads at high speeds on cross country routes, inevitably spelled doom for the Scenic Highway as a major commercial thoroughfare. Yet certain sections of it still remain in use for the tourists to see and admire and as a way of taking them past the many waterfalls in a leisurely fashion and offering them panoramic views through their camera lenses.

The two major sections open to the touring public today are the western, including Crown Point and extending a short distance east from Horsetail Falls, and the eastern, beginning at Mosier, between Hood River and The Dalles, and climbing to Rowena Crest with the winding descent down from it known as the Rowena Loops.

The view from Rowena Crest is magnificent, similar in scope to that from Crown Point, but different because the eastern end of the Gorge opens into the high desert country instead of the lush forest prevalent in the west.

Many short sections of the highway are visible from the I-84 freeway, sometimes seen only as a fleeting glimpse of a rock wall. One section a few miles long runs east from Hood River, although it is frequently closed by a gate. Hopefully, it will again be opened in the future for the view possibilities are among the best in the Gorge.

Some sections of the highway have been destroyed and still others are hidden away in the forests of the Gorge, half reclaimed by nature, with moss covering the warrentite paving and tree limbs arching across the top, closing out the open view of the sky.

Attempts are being made to inventory all of the roadway that remains so that it can be preserved, for at a time when the beauty and future of the Gorge is imperiled by those who would make it into a "commercial success" there is renewed interest in the old road.

The Columbia River Scenic Highway was, for at least a little while, perfection. As we drive along those few miles of it that remain we can marvel at the purity of the motives that built it and sense, here and there, the love and caring that went into it as we see the beauty of the bridges, or the elegant stone walls, or the wonderful symmetry of a curve that reveals a breathtaking view.

10

If we are today, in retrospect, astonished by the achievement, then that is nothing compared to the wonder voiced at the time it was opened. Let others who were present at the time of its first glory speak.

The editor of one of the great New York daily newspapers who said, editorially, after motoring through the Gorge, "The people of the Oregon Country have constructed perhaps the greatest highway, in the most magnificent setting in the world. They have pierced the mountains through with cloistered tunnels and carried the road around sheer precipices on buttressed walls, being careful to keep the natural beauty all about them and not to mar the landscape."

Frederick Villiers, British War Correspondent for the **Illustrated London News,** as he watched a sunset from Crown Point: "It possesses the best of all the great highways in the world, **glorified!** It is the king of roads."

Interestingly, although he was driven by great determination and presumably a strong ego, as are all successful persons, there is in Lancaster's many references to the Scenic Highway a humility, a willingness — nay, even an eagerness — to give credit to others for their assistance in the achievement.

And there is, also, a sense of history, perhaps best exemplified by his report of the national dedication of the highway, held on Wednesday, June 7th, 1916. After describing the ceremonies Lancaster concluded with the words, " 'It is toward evening, and the day is far spent,' let us work while we can for Our Country."

Simon Benson, a Portland businessman and believer in fountains on the city's street corners, had long advocated that a road be built through the Columbia River Gorge.

Samuel Hill was a Washingtonian who believed a highway through the Gorge could match the most famous roads in Europe.

John Yeon (left), shown here with another Portlander, J.W. Killingsworth, dedicated two years of his life, at one dollar per year, as roadmaster for the Scenic Highway.

Samuel Lancaster remembered his mother's admonition, "be careful of my Boston fern," and held destruction of the environment to a minimum as the highway was constructed.

SAMUEL C. LANCASTER
1864 - 1941
CHIEF ENGINEER
SCENIC COLUMBIA RIVER HIGHWAY 1913 - 1915

PIONEER BUILDER OF HARD-SURFACE ROADS. HIS
GENIUS OVERCAME TREMENDOUS OBSTACLES,
EXTENDING AND REPLACING THE EARLY TRAIL
THROUGH THE COLUMBIA RIVER GORGE WITH A
HIGHWAY OF POETRY AND DRAMA SO THAT MILLIONS
COULD ENJOY GOD'S SPECTACULAR CREATIONS.

Don Lowe

Plaque at Crown Point

I.

Samuel Christopher Lancaster was a marvelously complex man with enormous talent and an iron will. It seems safe to assume that one of his qualifications for building the Columbia River Scenic Highway was the fact that he had faced, in his personal life and at a relatively early age, great adversity and personal trauma which he had overcome through sheer determination.

He was born in Magnolia, Mississippi in 1864, but the family moved to Tennessee when he was a small boy. His father lost most of his money in one of the financial panics and, as the eldest of five children, it was necessary for Samuel to assume many of the family responsibilities. He finished one year of college and then, under the guidance of the chief engineer of the southern lines of the Illinois Central Railroad, he continued his engineering studies.

While doing railway construction work at the Yazoo Delta he contracted malaria and it was believed that this disease, combined with the long hours he had been working, contributed to his being felled by infantile paralysis, which incapacitated him for quite some time.

"I lay in my bed at my home in Jackson, Tennessee, for eighteen months, practically paralyzed," Lancaster told an interviewer later in his life. "I could move my head a trifle, but that was all."

A doctor, convinced the paralysis was complete, stuck needles under Lancaster's nails to confirm that the nerves were deadened. Lancaster howled with pain because the nerves were even more sensitive than they had been before.

Still his fingers, toes and limbs were crippled and bent entirely out of shape and gradually the tendons began to harden. Haunted by a tale he had heard of an "ossified man" used as an exhibit in sideshows, and whose disease had begun in a manner similar to his, Lancaster vowed to emancipate himself from complete helplessness. In his efforts he was aided greatly by his mother and the girl who was to become his wife when he recovered.

He began his recovery by concentrating on moving the middle finger of one hand. About this time his mother read to him a true story about a boy, crippled in the same way who, unable to move either hands or feet, had learned to paint pictures with a paintbrush held between his teeth.

It took a courageous effort, but Lancaster managed to get a pencil between his teeth and then, by shifting his weight in bed, he was able to train himself to make a few crude letters, and eventually to write coherent messages.

Determined to walk again, despite his doctor's prediction that he would not, he persuaded members of his family to lift him upright, but he collapsed "like an empty potato sack" when they loosened their grip.

His brain was still clear and sharp and using the pencil between his teeth he laboriously designed a frame into which he could be strapped, suspended in air. The pain was intense, and he could endure it for only a few minutes at a time, but each day he practiced until one day he managed to stretch his legs enough to touch the floor and he was then able to move the frame on its ball-bearing casters. He moved around the house in this manner for a time, but resolutely decided he must stand alone, without support of the frame, which he finally succeeded in doing for a few seconds before he crashed to the floor.

Despite the excruciating pain he discovered that the brief period his weight had been on his toes had caused the crippled tendons to be bent from their unnatural position, so he next set about the grim business of loosening all his stiffened tendons through massage and exercise. He graduated from the frame to crutches, although his fingers remained so stiff that he could not grip a pencil. At this point some city officials came to him for help on an engineering problem and although he had not been able to work for two years he somehow was able to do the required drawings.

Although still crippled, he was hired by a local railroad and as his health improved and his strength was restored he performed so many services for the city that a public park in Jackson was named in his honor.

In 1904, Lancaster became a consulting engineer after the Secretary of Agriculture became intrigued by a system of country roads he had designed in Tennessee, and in 1906 he moved to Seattle where he became acquainted with Samuel Hill.

What makes Lancaster fascinating is the strange blend of practicality and romance in his personality; the stubborn resolve that enabled him to recover from a debilitating illness and the engineer's mathematical mind mixed oddly with the tender soul of a poet who appreciated nature and all its glories.

"My love for the beautiful is inherited from my mother," he told another interviewer after the Scenic Highway was completed. "When I made my preliminary survey here and found myself standing waist-deep in ferns, I remembered my mother's long ago warning, 'Oh, Samuel, do be careful of my Boston fern.'

"And I then pledged myself that none of this wild beauty should be marred where it could be prevented. The highway was so built that not one tree was felled, not one fern crushed, unnecessarily."

In 1908 Samuel Hill invited Lancaster to accompany him to Europe to study some of the highways on the continent, which journey Lancaster recalled in this charming "Reminiscent" in his book.

The Rhine has been a favorite resort for lovers of natural beauty for more than a thousand years. In the fall of 1908 Mr. Samuel Hill and I were delegates to the First International Road Congress at Paris, and were attracted to the Rhine while studying road conditions throughout Continental Europe. Shortly after passing Bingen we succeeded in making excellent photographs of some of the old castles. The ancient ruins of Ehrenfels, perched high on the steep slopes of the Rudesheimer Berg, charmed us, for the hillside at that point is completely covered with terraced vineyards, supported by massive masonry walls resembling giant stairs.

"Who built these walls?" was the question asked.

"Charlemagne," was the reply, — "And you are going to see something like that on the Columbia some day."

"What — Castles and Vineyards?"

"No," said Mr. Hill, "I am thinking of a great highway — although you may see the castles and vineyards, too."

In my heart I said, "Not in my lifetime — If I ever have a grandson, he may live to see it, or, maybe his son."

Only five years elapsed, however, when, through Mr. Hills influence, the author was asked to fix the location and direct the construction of this highway through The Gorge of the Columbia.

Dry masonry walls — (constructed without the use of cement, or mortar of any kind) have been used extensively in Germany, Italy, Switzerland and Greece in bygone centuries.

We adopted this plan and constructed many miles of dry masonry walls on the steep slopes of the mountains in the Columbia Gorge. They add greatly to the charm of the highway.

The Italian laborers built their very souls into these walls as they sang their native songs and thought of the homeland.

Samuel Christopher Lancaster
September, 1926.

Lancaster remained grateful to Samuel Hill throughout his lifetime for granting him the great opportunity to build a road

through the Columbia Gorge. In an earlier edition of his book one page is headed

Samuel Hill

Road Builder

and beneath this heading Lancaster had written the following:

Who loves this country and brought me to it
Who showed me the German Rhine and Continental Europe.
Whose kindness made it possible for me to have a part in planning and constructing this great highway.
There is a time and place for every man to act his part in life's drama and build according to his ideals.

God shaped these great mountains round about us, and lifted up these mighty domes into a region of perpetual snow.

He fashioned the course of the Columbia, fixed the course of the broad river, and caused the crystal streams both small and great to leap down from the crags and sing their never ending songs of joy.

Then He planted a garden, men came and built a beautiful city close by this wonderland. To some He gave great wealth — to every man his talent, — and when the time had come for men to break down the mountain barriers, construct a great highway of commerce, and utilize the beautiful, which is "as useful as the useful," He set them to the task and gave to each his place.

I am thankful to God and His goodness in permitting me to have a part in building this broad thoroughfare as a frame to the beautiful picture which He created.

Samuel Christopher Lancaster

Highway Engineer

1915

In modern day terms Lancaster's words have an archaic, almost Biblical ring, yet there is reflected in them the grandeur he felt in the Gorge of the Columbia. Consider the following, also from his book.

Don Lowe

Lancaster loved the great basaltic cliffs that tower above the Columbia. This is Rock of Ages.

Formation of the Cascade
and Sierra Nevada Ranges

There was a time when the waves of a nameless ocean kissed the Western slopes of the Rocky Mountains — when unborn continents lay still in the dark, cold womb of fathomless seas. Even then, far — far off shore, the voice of God was heard, and out of the boundless deep He lifted up a mountain range. From North to South it rose like some leviathan stretched at full length, with head and tail touching the mainland, and the Cascade and Sierra Nevada Ranges were created, thus forming an inland sea, a thousand miles or so in length.

How fearful were the sounds! How dark the skies! The earth groaned and trembled as if in travail when this new land was born; the very foundations were broken up, and flames burst forth. The rocks were melted with fervent heat, and white hot magma streams ran down the mountain side into the sea. Steam rose in clouds — lightnings flashed — rain poured in torrents — thunder roared. The whole mass heaved, and rose, and fell, as a bosom moved with passion, until that day's work was done.

When the sun broke through the veil, it shone on a naked land, its only clothing ashes — hot ashes — blowing, drifting everywhere.

For centuries the most active volcanoes were at work. They built up mighty domes reaching into the skies, one mile, two miles, almost three miles high, until the icy-cold of the atmosphere, where they now reared their heads, exceeded the cold of ocean depths whence the uplift came.

Time first closed the smaller vents and fissures, then hushed the greater ones. When the fires from within were extinguished, perpetual snow crowned the loftier peaks.

These great snow fields moved slowly, sliding, pushing downward, producing many an avalanche, and glaciers extended far into the lower valleys.

In their imperceptible movements these mighty glaciers wore through the lava beds in many places, cutting gashes hundreds of feet in depth, grinding to powder the older limestone and other rocks beneath them. These fragments were mingled by the hand of God with ash and other particles of the igneous mass, which He took from the very bowels of the earth. The little rivulets joined with mountain torrents to bring the product of the glaciers down into the valleys, where He spread it out, producing a soil rich in everything that ministers to man.

Then the Prince of All Gardens planted the seeds of a thousand springtimes. Some flowers He made to grow high up in the clefts of the rocks, in fields of snow. The anemone and heather He planted a little lower down, just where the trees begin; and when He came to where the earth slopes gently out in upland meadows, jeweled with sparkling waterbrooks, He gave more freely of His abundance and carpeted the earth with flowers of every tint and hue. Alpine firs He planted here and there, grouping them, and adding others as He came down into the valleys, where He made the flowering shrubs and ferns to grow in the midst of dark, cool forests of great and stately trees, the shelter of His creatures.

There is a beauty in the bare angles of the rocks which look down from the heights, where His fingers broke them. Here He rent and tore them asunder, to make room for one of earth's great rivers.

This, then, was the man selected to design and construct the Columbia River Scenic Highway. It was to be his ultimate achievement, his memorial.

From the very beginning it was essential to Samuel Christopher Lancaster that the Gorge and the road he built through it remain in harmony, that they complement each other in the same manner as the reflected glow that is on the faces of lovers when they are together.

In this sense, he was building more than a highway. He was creating a legend.

Oneonta Gorge

Lancaster, shown here in center (profile) enjoyed the Oregon outdoors. This picture is on Larch Mountain summit.

Don Lowe

Earlier efforts to get a road past the barrier of Shellrock Mountain failed. This is remains of road building attempt prior to the construction of the Scenic Highway.

II

The Columbia River was the last great river to be discovered and is the only water level route through the Cascade Range between Canada and northern California.

Thus the vast majority of early settlers poured into western Oregon through this gap in the Cascades, although some overland routes through the mountains were also established, but sparsely travelled because of the difficulties involved.

Agitation for a road paralleling the Columbia was begun only a few decades after the river was discovered. The first wagon road on the Oregon side was completed on February 9, 1856. Less than six miles long, it ran from Bonneville to Cascade Locks and climbed to an elevation of more than four hundred feet to get over a point of rock, at the base of which ran the Portage Railroad.

No one seems to be certain who was responsible for building the road, but on February 9, 1856, an article in **The Oregonian** reads: "We are informed that a new road around the portage of the Cascades, on the Oregon side, has been completed and that goods are now being transported over this road with safety and dispatch."

On the same date, another article in **The Oregonian** states that W.R. Kilborn, who lived at the Lower Cascades, had "per-

29

fected arrangements for transportation of freight over the portage at the Cascades on the Oregon side," and quoted Kilborn as saying, "The road is now in complete order and my teams will always be in readiness."

On March 23, 1861, Brigadier General Rufus Ingalls, in a letter to J.W. Nesmith, stated: "I have always felt deeply impressed with the necessity of having a good wagon road from Vancouver to The Dalles, probably passing the Cascade Mountains on the Oregon side of the Columbia."

Three years later General Ingalls again stressed the need for a road, remarking: " The demand for a land route through the Cascade Mountains becomes more serious and important every day. As a military measure, it is important to connect the lower Columbia with the great interior by a practicable wagon road. I have seen the importance of it during the Indian wars. It will be still more necessary in case of foreign wars."

On October 23, 1872, the Oregon State Legislature appropriated $50,000 to build a wagon road from the mouth of the Sandy River through the Gorge to The Dalles. In October, 1876, another $50,000 for the project was provided by the legislature.

The end result was a narrow, crooked road with extremely steep grades and much of it was destroyed in 1882-83 when the railroad was built through the Gorge.

In 1910, E. Henry Wemme, an early automobile enthusiast in Portland, circulated a petition to have a road built from Bridal Veil Falls east to the Hood River County line, and in 1912, following a survey, almost two miles were built. But, it had to be relocated in 1915 because the road did not conform to state trunk road standards established by the newly formed State Highway Commission.

In 1912, Oregon Governor Oswald West, who had been experimenting in southern Oregon with convict honor camp labor in road building efforts, was the recipient of a gift of $10,000 from Simon Benson, the money to be used for the construction of a road around Shellrock Mountain in Hood River County. This mountain had long been considered an impassable barrier and although this latest attempt was not successful it again called attention to the need for a road through the Gorge.

Also in 1912 Samuel Hill made a walking trip with some

friends over the area between Warrendale and Eagle Creek and talked to them about the possibility of a great highway that would someday extend along the south side of the Columbia.

Roads were a passion in Hill's life. In the state of Washington, where he lived, he agitated for a north-south cross state highway. He founded the Washington State Good Roads Association, and he also spent a considerable amount of time and money advocating the Alaska Highway. At one point in his lifetime the Good Roads Association people wanted him to run for the United States Senate but he refused, believing that he would be better remembered as a road builder than as a politician.

In his European travels he had been fascinated by the great roads on the continent. The famed Axenstrasse, in Switzerland, with its three windows overlooking Lake Uri, appealed to him particularly.

Samuel Hill may not have been the catalyst which caused the Scenic Highway to be created — it is probable that events in the world and pressures of civilization were the main cause — but certainly he must be given credit for seeing to it that it was a singular road, as distinct from the ordinary as Hill was distinct from the ordinary man.

Hill was born in North Carolina in 1857, son of a well-to-do physician, who, because of his strong Union sympathies, moved to Minneapolis at the beginning of the Civil War.

Ten years old when his father died, young Sam worked at odd jobs, saved his money and eventually went to Harvard where he acquired a degree in law. It is believed that at some point in his career he attracted the attention of James J. Hill, the railroad magnate (no relation). Sam went to work for the railroad and, in 1888, married James Hill's daughter.

At a time when wheeler-dealing was as commonplace as heading west to seek your fortune, Sam Hill made a lot of money. Some have speculated on the source of his wealth, but no matter. Unlike other entrepreneurs of that age he was never caught up in scandal.

European royalty fascinated him also, and it was said he started to build his Maryhill estate (named for his daughter) as a place to entertain his friend, King Albert of Belgium, who planned to visit America in 1914. (The war prevented the king's visit and in later years Hill had Queen Marie, of Rumania, come to the United States and dedicate Maryhill as a museum. Hill also built a replica of Stonehenge east of Maryhill, and among his other monuments is the Peace Arch at Blaine, Washington, on the Canadian border, which he planned, built and financed).

Hill's 1912 walking trip in the Gorge was no idle stroll, but part of his plan to have a road created in the grand concept he envisioned. A year later the Oregon State Highway Department was organized and in February, 1913, Hill had as his guests (via a special train to his Maryhill estate) the members of the Oregon Legislature.

He had built several miles of hard-surfaced road at his own expense near his ranch and he had the members inspect not only those roads but also sit through an illustrated lecture on the subject of road building. As an outgrowth of this the legislators enacted a law creating the position of State Highway Engineer.

Initial budget for the entire department was $10,000, which was set aside for office expenses. An educational program was begun to teach Oregon counties how to vote bonds to build trunk roads and in short order the counties of Clatsop, Columbia, Multnomah and Hood River raised a total of $2,235,000 for the Columbia Highway.

Today we can only marvel at the speed with which all of this was accomplished. In a modern bureaucratic world it would be impossible.

Equally impressive is the unanimity of opinion on the type of road that must be built. These road builders loved the Gorge and took great pains to avoid damaging its scenic values in any way. Certainly a far cry from later generations who pushed the completed project (whatever it was) ahead of considerations for preservation of natural beauty.

Perhaps Sam Hill, when looking at the Rhine, was reminded of his own view of the Columbia from his several thousand acre ranch on the barren plateau above the Columbia's north shore and felt a kinship between the ancient European river and the Great River of the West on the raw American frontier.

Hill's particular genius and his great contribution to the Columbia River Scenic Highway was to somehow transmit his vision to the one mind — Samuel Christopher Lancaster's — who had the skill and talent to make it into a reality.

Thus, by 1913, the grand passion of Lancaster's life was beginning to unfold. Only five years earlier he had listened in disbelief as the likable but admittedly eccentric Samuel Hill had shown him the ruins of Ehrenfels and told him that someday there would be a marvelous highway through the Gorge of the Columbia. Now construction on that road was started in October.

Don Lowe

Samuel Hill built Maryhill in anticipation of a visit by his friend, King Albert of Belgium.

Interstate 84 from a section of the Scenic Highway now inacessible by car. Rooster Rock sits between the freeway and the Columbia River.

III

The modern day Gorge traveller, zipping along Interstate 84, might have some problem understanding the difficulties encountered when construction began on the Scenic Highway.

Some reminders are in order. In 1913 there were no dams to control the flow of the Columbia, the level land close to the foot of the cliffs had been appropriated as railroad right-of-way, and old photographs reveal that the highway was built with the tools of the times — manpower, mulepower and some steam engines, instead of the mammoth earth moving machines we take for granted today.

In addition, there were those special considerations with which the builders were charged. A road with a minimum width of twenty-four feet, with extra width on all curves, no radius less than one hundred feet and a maximum grade of five percent.

The last named was, in itself, a staggering problem, as grades on previous Gorge roads had seldom been less than ten percent and had been as high as twenty percent.

Specifications aside, there remained the one overwhelming consideration. This was to be more than a highway over which automobiles could race to the opposite side of the state. It was meant to be more than a convenient passage. The scenic highway designation required a design to fit in with the natural beauty of the Gorge, in such a manner that panoramic views were available for the motorist, as well as comfortable spots along the way to pause and enjoy such roadside attractions as the waterfalls that crowded the western section of the road.

In view of these facts, it seems that destiny did, indeed, call Samuel Christopher Lancaster to the task. The obstacles that needed to be overcome were tremendous and what better person to meet the challenge than the man who had fought and won one of life's great battles, who had stood erect and walked unaided again after his doctors had told him he was forever consigned to physical inactivity?

In building the Scenic Highway there were four major problems and out of these emerged four distinct engineering feats of skill. These were; the bridges, the tunnels, the viaducts and the road across Crown Point.

The early going was easy enough. East from Portland the road ran mostly through meadows or climbed open grades until it was on top of the crest at Chanticleer Point, today's site of the Women's Forum Park.

Trouble lay directly ahead. The road must now descend to Latourell Falls, near water level, and Crown Point, towering more than seven hundred feet above the valley floor, barred the way. In 1910 the county surveyor had said a road from Chanticleer to Latourell was impractical, even allowing for a maximum grade of twenty percent. Some thought was given now to have the road skirt the base of the rocky cliff, but in order to do that much of the rock would have to be cut away and the grades would still be steep and the curves too sharp.

Lancaster had already clambered laboriously to the top, inspecting the obstacle from the bottom up, and he came away with what was considered to be a daring solution. According to one story his associates chorused "impossible!" when he proposed it. And, it is said, he smiled in reply, and it was done.

The decision was to go over the top and down the far side. The problem, simply put, was that if the specifications involving the radius of turns were followed, any attempt to do so was doomed to failure for the road would not only run to the very rim of the precipice it would, in spots, literally go off into thin air.

Lancaster's solution was to build concrete supports up from

Section of the Scenic Highway leading from Chanticleer Point to Crown Point.

the broader shelves of rock below. In the descent it was necessary for the highway to parallel itself five times in a series of graceful turns in a very short distance.

This section of the road still exists and is greatly admired and quite heavily travelled, particularly in the summer months. On the road leading to Crown Point from the Women's Forum Park the first rock walls along the side of the road become visible and one begins to sense this is more than just another highway. Then there is the panorama from Crown Point itself, where the Vista House was built and still stands, like some ancient castle guarding the gateway to the Gorge below.

It is almost three miles to Latourell Falls, first of the major Gorge waterfalls and from there on the falls come tumbling, bouncing, roaring down the canyon walls, sometimes free falling hundreds of feet, as in the case of Multnomah Falls. Each stream of any size requires a bridge to cross it.

On the Scenic Highway each bridge was designed individually, with care and attention to detail. The bridge at Latourell is 240 feet long with three 80-foot spans. It was designed by J. Billnes, and today one can park after crossing it and walk down to the base of the falls via a pathway.

A bit farther on is Shepperd's Dell, where George Shepperd gave eleven acres of land, and a beautiful, curving walkway leads down to the base of the first falls while the second falls is crossed by a single 170-foot arch span, 130 feet above the racing water.

At Bridal Veil Falls a 110-foot long bridge crosses directly over the top of the falls, making it the only one not easily visible from the roadway. At Wahkeena Falls a parking lot is provided and across the road a park and picnic area.

Multnomah Falls, largest by far of the falls in the Gorge and second highest in the United States, is the most popular stopping point in the Columbia. Multnomah Creek is crossed by an arch bridge while from the highway can be seen a footbridge which crosses the creek between the upper and lower sections of the cataract.

Shortly before Mulnomah Falls and for some distance beyond it the road is carried on a viaduct, a system that was to be used in other sections as well. All viaducts are 20-feet wide on columns

Crown Point as seen from Chanticleer Point, with Beacon Rock in background on the Washington side of the Columbia.

Shepperd's Dell. Photo was taken in either 1915 or 1916.

Bridge across the Scenic Highway near Latourell (known then as "Talbot's Place.")

Bishop's Cap was the East Dome at Shepperd's Dell.

set 20-feet apart in rows of 17.5 feet apart and it is a testimonial to their original construction that they are still in use today.

Near Oneonta Gorge, which is slightly more than two miles east of Multnomah Falls, a several hundred foot high cliff of solid rock, with the railroad taking up all the available space at its base, necessitated creation of a tunnel more than 100 feet long. It was the first of several tunnels along the route, although none of them remain open to traffic today, having been either closed or one side blown off to open the roadway.

The problem at the Oneonta Tunnel was that in order to prevent thousands of tons of rocks from cascading down onto the railroad tracks when the blasting began it was necessary to go to considerable extra work to strengthen the cliff before digging into it. So the weaker sections were plugged with concrete before the blasting started, one of Lancaster's many innovations.

A 60-foot bridge spans the creek at Horsetail Falls, just east of Oneonta. The Falls are located so close to the highway they spread mist across it constantly. The next creek up the Gorge, McCord, was crossed by a 360-foot long bridge.

The Moffett Creek Arch, a short distance beyond McCord Creek, still stands today, directly alongside I-84, westbound. Although it is no longer in use it should be preserved for it was one of the engineering marvels of its day, the longest three-hinge re-inforced concrete arch bridge that had ever been constructed. It had a clear span of one hundred and seventy feet and the "flat arch" rose only one foot for every ten feet in length. The hinges were of massive cast iron with 4 1/2-inch steel pins, and it was designed to carry a live load up to 200,000 pounds, distributed uniformly over half the span.

Tanner Creek was crossed on a 60-foot concrete deck girder bridge and the unique Eagle Creek, one of the most beautiful bridges along the highway, has a full-centered arch with a span of 60-feet and overall length of 144-feet. Its re-inforced concrete is faced with native stone.

East of Eagle Creek, and reached easily by walking along a trail from Eagle Creek Park, a long deserted section of the Scenic Highway remains, beginning with a small stone bridge over Ruckel Creek. Ruckel Creek falls in two drops, the first one

Oneonta Tunnel

The Moffett Creek Arch was hailed as an engineering master-piece.

Bridge over Ruckel Creek can be reached by taking the Ruckel Creek Trail from Eagle Creek Park.

immediately south of the bridge, the second immediately below it. From the bridge the road wends its way eastward, covered now with moss and sometimes partially blocked by fallen trees or tree limbs. It is a pleasant walk to where a fence, overlooking I-84, bars any farther advance. There has been some discussion of creating a picnic/park area here, but to date nothing has been done about it. It is worth seeing and worth saving.

Previous road building attempts had come to grief at Shellrock Mountain which is, basically, a conical pile of sliding stone, named for its resemblence to shelled corn. The lower slopes reached almost to the waters edge and it was necessary for constant maintenance then and is even today on I-84.

Mitchell Point looms seven miles past Shellrock Mountain and it was here Lancaster met his second barrier and turned it into triumph, perhaps the greatest along the entire route.

In his conversations and in his writings Lancaster refers repeatedly to the Axenstrasse, the Swiss highway, and its three windows overlooking the lake with the views towards the mountains beyond.

Samuel Hill had taken him to Europe and had shown him the Axenstrasse and surely Lancaster, when he viewed Mitchell Point, recognized the marvelous opporunity to excel that legendary Europe feat of road building. So here was created the "Tunnel of Many Vistas," with no less than five huge windows cut in the solid rock, affording a view across the river to the Washington side.

It was no easy task. In some places the cliff actually overhung the river and engineers were lowered on ropes as much as 200 feet where they hung like spiders over the torrent as they planned their moves.

Again, great care had to be taken in blasting as the railroad hugged the base of the cliff far below. In the end it was a masterful design, a 390-foot long, 19-foot high tunnel which, at one point, came within 10 feet of the face of the cliff in order to allow for the prescribed curve radius. It is doubtful if any other feature of the Scenic Highway caused so much national comment or won so much praise.

In his writings Lancaster gives credit to J.A. Elliott, working under the supervision of Major H.L. Bowlby, State Highway

Engineer, for the Mitchell Point Tunnel. In some other places it is referred to as "the Bowlby Tunnel." Yet it would be difficult to imagine Lancaster not having considerable say in the matter, particularly in view of his and Samuel Hill's interest in the road that had served as, if not their model, at least their inspiration.

Regrettably, the Tunnel of Many Vistas at Mitchell Point no longer exists. Some say it was blasted shut for safety's sake and others claim its destruction was a travesty and a senseless waste.

The Scenic Highway was opened officially between Portland and Hood River on July 6, 1915, and the extension to the Pacific Coast was opened on August 11th of the same year.

One can imagine the sensations experienced by the early travellers. Rather obviously, it signalled for Oregonians the beginning of the age of the automobile, giving them a new freedom.

Ironically, it was the automobile and the rapid advances in its technology, that doomed the highway. For the road was built by romanticists in a romantic time. Practical men they undoubtedly were, but dreamers at heart, still entranced by the magic of lovely scenery, still feeling the mystery of creation and standing somewhat in awe of it as they approached their task.

Their purpose was not to conquer, but to enhance. What they built was a Camelot, an ideal. It was, without question, a road right for its time, but that time was so short, so fleeting, that progress rushed by it without pausing hardly to glance at their creation.

So what remains of Lancaster's Road are remnants of a vision shared by some forward looking people. In the words of the Consulting Engineer:

> For the first time in history it is possible to drive a wagon from the wheat fields of Eastern Oregon through the Cascade Mountains to the sea, and to those who have always thought with Sir George Simpson that it could never be done, we paraphrase the answer of Doctor Marcus Whitman and say, "There is a wagon road through these mountains, for we have made it." An automobile can cover the entire distance in one short day's travel, and no man

Mitchell Point Tunnel was considered a work of art.

can estimate the value of this great Highway to all the people of the Pacific Coast.

Men of all stations have vied with each other in their giving, to help along this work.

Mr. John B. Yeon contributed two years of active service, aided by his friend and neighbor, A. S. Benson.

At Crown Point approximately two acres were donated, through the kindness of Osmon Royal.

George Shepperd gave eleven acres for the good of his fellows, Simon Benson made it possible for thousands to enjoy a broad river, high mountains, and sparkling water falls, by his gift of more than three hundred acres.

Jacob Kanzler found a way for our National Government to assist by setting aside some fourteen thousand acres of National Forest for the free use of all, in which tired men and women with their little children may enjoy the wild beauty of nature's art gallery, and recreate themselves.

Just two years ago there came together a small group of men, who were fully assured that the time was ripe for the inception of this great work. They set about the task with strong determination, and brought it to pass......"

From these men came the paen of praise, the love song, the poet's dream, that was the Columbia River Scenic Highway. They were inspired.

As Lancaster said:

All who enter the Columbia Gorge and look at those great monoliths that lift their heads high above the clouds, where this Continental River has been at work through long centuries of millenniums, carving its way through the very heart of the mountain range, cutting the hard rock down to the level of the sea, will be impressed by an observation made by Thomas Jefferson more than a hundred years ago when he said, "It is a solemn and touching reflection perpetually recurring, of the weakness and insignificance of man, that, while his generations pass away into oblivion, with all their toil and ambitions, nature holds on her unswerving course, and pours out her streams and renews her forests with undecaying activity, regardless of the fate of her proud and perishable sovereigns."

"Tunnel of Many Vistas," Mitchell Point

Don Lowe

Mist Falls, seen here from Interstate 84, is one of many waterfalls that hang from the basaltic cliffs of the Columbia River Gorge.

epilogue

On a late January Wednesday, when the air was filled with the cold gloom of winter and dark clouds came to sit only a few feet above the snow line along the high ridges of the Gorge, I took a business trip from Portland to The Dalles.

I had learned, long ago, that sentimenal journeys pose inherent dangers and so had deliberately avoided such a trip, knowing full well that he who retreats into the past can easily become so overcome by nostalgia that he is entrapped and bewitched by it. Unabashed old Gorge lovers are, after all, subject to fits of melancholia as they view the "progress" and a writer must retain at least some semblance of objectivity if he is to remain credible.

In an average year I drive through the Gorge no less than a dozen times on business and I admit readily that Interstate 84, an almost level route which allows me to cruise along at the maximum allowable speed, serves my business purposes better than the Scenic Highway ever could. And, reluctantly perhaps, I admit also that even from the freeway the broad river and the dark, basaltic walls still exert their charm.

Because of my trips I see the Gorge in all weathers. For some, I am sure, the best time is on those clear days of summer or fall,

51

when the view is unimpeded by lowering clouds and fog or rain, when the sun splashes the gloomy walls with brilliant light and the sky is a glorious shade of blue, with perhaps a small white cloud for accent, forming a perfect background to the deep green of the firs.

Yet, for me, winter has always been the perfect, most expressive time to be in the Gorge. For it is then that a hundred unnamed waterfalls spring to life and hang from the great cliffs in tiny, silver threads, to vanish when the warm sun dries the uplands and only the familiar water channels perform their full year function.

In the winter there are also those small but raging torrents visible on the Washington side, coursing their white way down the hillsides of the north shore. These streams, too, will retreat with the springtime sun.

I have never understood why those misty, brooding days have always been my favorite time in the Gorge, for I am given more to sunlight and dry air in other areas.

No matter. This day was a business trip and even my recent research into the life of Samuel Christopher Lancaster, Sam Hill and the rest of those roadbuilders was not going to cause me to stop on my appointed rounds nor get me to swerve from the path of rigorous duty.

So it was with a sense of surprise that I found myself turning off Interstate-84 at Mosier and heading through the town and then up through the orchards toward Rowena Crest. A great deal of building is going on up there now and I was intrigued by the interesting shapes of some of the residences-to-be. Then as the road leveled out there was ahead of me a solid wall of white rising up from the Columbia River valley. I had climbed literally into the low clouds.

As I turned off to drive the loop to Rowena Crest Viewpoint I passed a black Porsche parked along the side, with two lovers locked in their passion and oblivious to my passing, and as I stopped my own car and got out of it the white veil in front of me parted and I saw again the wondrous view upriver from this lofty aerie.

Until that moment I had been consciously avoiding sentimentality. Even my little side trip away from the freeway was not

52

The Rowena Loops wind gracefully down from Rowena Crest.

going to cause me to depart from my day's schedule. But in the moment when the mist parted and the valley was revealed, the Circe who inhabits the Gorge called to me as she called to others in ages past and I heard the siren song that not only brings persons back to view it again and again but imbues them with a kind of mystic wonder at the marvels of creation, for the entire Gorge is that kind of cathedral experience.

There are in this world places of great beauty, where only poets and musicians hear nature's call, but in the Gorge of the Columbia it has been heard by hard headed and practical men in days long gone......Samuel Hill, John Yeon, Simon Benson.... the list goes on through generations past and present, including former Oregon Governor Tom McCall who, as late as April, 1981, expressed fears for the future of the Gorge if plans then considered were carried to their conclusion.

So maybe all who love the Gorge had damn well better forget about **avoiding** sentimentality, and accept it instead. This is a love affair, and love requires no apology.

Samuel Christopher Lancaster made a unique contribution when he created a road of breathless beauty, even as he acknowledged his efforts were but "the work of pigmies."

Perhaps Lancaster knew his road, in its entirety, was to last only for a little while. Certainly he knew that he had done the best he possibly could and when it was finished he said: "When all plans had been fully completed, and practically all of the work of construction had been finished, the author's official connection with it was severed, but his affection and interest in the work will last throughout the remaining years of his life."

On this January day the sun dispersed the clouds on Rowena Crest and the impounded river ran like a wide blue ribbon through a green valley. Behind me the black Porsche stirred to life and the lovers roared away to seek another trysting place.

Alone I stood at the rock wall, looked down from it at Lancaster's Road twisting its way back to the valley floor, and said to myself, let business wait for just a little while, man is not given an opporunity to savor such a view every day.

View east from Rowena Crest.

The Columbia River Scenic Highway was a labor of love, for those who created the plan for it, but real labor for those men and animals who put in long hours daily to insure its completion. Many groups volunteered to spend a day on construction, but the work was continued by the regular crews long days after the volunteers went back to their regular jobs.

Construction on the Scenic Highway, April 25, 1914.

Volunteers pitched in to get the project underway. These are members of the Portland Ad Club and Realty Board.

Men, mules and machines combine their efforts to build the road.

Many of the scenes shown on the following pages exist only in these old photographs or in the minds of those who travelled th Scenic Highway before it was supplanted by straighter, but less enticing, roads.

Through these photos we take a sentimental journey along the King of Roads as it was then, the dream realized for Samuel Hill, Simon Benson, John Yeon, Samuel Christopher Lancaster and all the others who worked to create it.

Vista House at Crown Point

Scenic Highway curves through the woods east of Crown Point.

Wahkeena Falls. Legend has it that Wahkeena is Indian for "most beautiful."

Multnomah Falls. Note footbridge above the lower falls.

Viaduct east of Multnomah Falls.

Samuel Lancaster (in rear seat of auto) enjoys ride along the Scenic Highway.

Horsetail Falls

Viaduct near Eagle Creek

Eagle Creek bridge is faced with native stone.

Section of the highway between Cascade Locks and Hood River

But it wasn't always so easy to travel, as this 1921 winter photo
reveals.

Another scenic view from the highway. Note fish wheel (later outlawed) in river.

The Scenic Highway successfully crossed Shell Rock Mountain, although earlier efforts had failed.

Highway hugs cliff above river as it approaches Mitchell Point.

Then plunges through the rocks to the tunnel entrance.

The "Tunnel of Many Vistas," as seen looking up from the railroad tracks on the river bank.

Deschutes Tunnel was one of many no longer in existence.

Cover Photo
  Oral Bullard
Abandoned section of the Scenic Highway
near Ruckel Creek.